DATE DUE			

11934

629.4
BER

Berliner, Don.

Our future in space.

Our Future in
SPACE

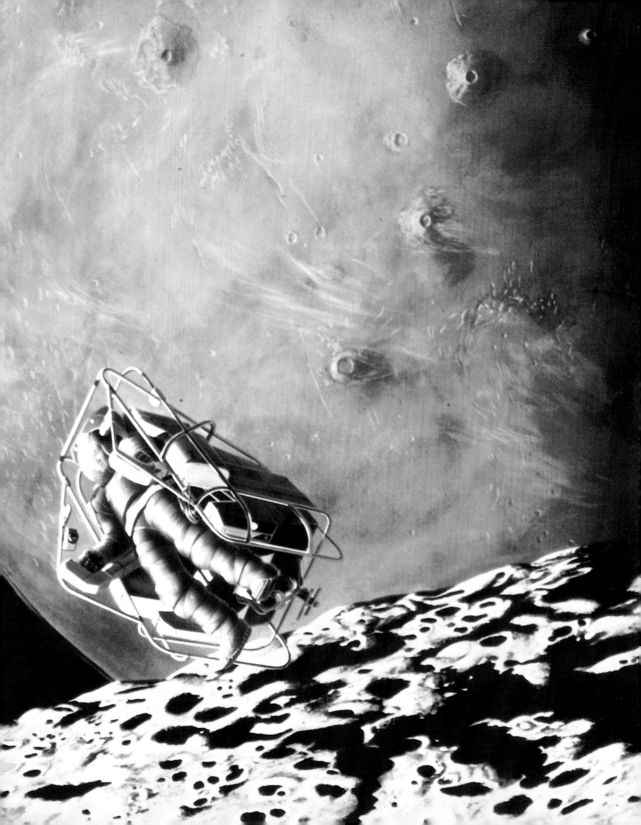

Our Future in

SPACE

Don Berliner

Lerner Publications Company • Minneapolis

Library of Congress Cataloging-in-Publication Data

Berliner, Don.
    Our future in space / Don Berliner.
        p.   cm.
    Includes bibliographical references and index.
    Summary: Examines current space research and future space projects
such as a permanent space station, an observatory and research base
on the moon, and the search for extraterrestrial life.
    ISBN 0-8225-1592-X
    1. Outer space—Research—Juvenile literature. 2. Space stations—
Juvenile literature. 3. Life on other planets—Juvenile
literature. [1. Outer space—Exploration.] I. Title.
QB500.262.B47   1991   90-13522
                                                                    CIP
                                                                     AC

Manufactured in the United States of America

2   3   4   5   6   7   8   9   10   P/JR   00   99   98   97   96   95   94

Contents

In this artist's concept of Mars, astronauts, lower right, *explore the planet's surface.*

Introduction

A dream can be about anything. It need not be practical or sensible. It need not even be possible, for it is only an idea. For centuries people watched the birds and dreamed of some day being able to fly in the air. Now, as people gaze at the stars and planets, their dreams are often about living and working in space, about trips to Mars and to planets even farther away, and about meeting those beings who may already live out there.

It wasn't until the late 18th century that humans were able to realize the dream of flying in the air. In 1783 a hot-air balloon designed by the Montgolfier brothers lifted off from Paris, France, with two passengers aboard. It took humans on their first journey through the sky. Less than 200 years later, in 1961, Soviet cosmonaut Yuri Gagarin fulfilled the other dream when he made the first journey into space. In a space-ship called *Vostok I*, Gagarin blasted into the darkness of space and orbited the Earth.

Without dreams, people would still be tromping across fields and over hills, never more than a few miles from home. But because of human curiosity, each time a person realizes one dream, at least one more quickly takes its place. In 1969 three men traveled 240,000 miles (386, 160 kilometers) to the Moon. Humans immediately began dreaming of a flight to Mars, more than 35 million miles (56 million km) away.

Dreams are only the beginning of every space project. If a dream is to become a reality, it must stand a fair chance of success, and it must involve serious planning. Planning must then be followed by a large investment of money, because space experiments and trips cost far more than any balloon or airplane experiments ever did. The cost of a new space project must be weighed against its possible benefits: Will it be more worthwhile than spending the money on Earthbound scientific research? Will it help humankind more than spending the money on pollution control, low-cost housing, or the war on drugs?

Almost no space projects offer a quick "payoff." It may take many years of hard work to produce any results of obvious value. Some projects, such as the Project Apollo landings on the Moon, may never have a direct impact. Of course, when Christopher Columbus set sail from Spain in 1492 and discovered the Americas, the effect of his discovery on the entire world wasn't understood until many years later.

We can look back on the first few decades of the space program and see some benefits: development of new materials, advance warning of hurricanes, and verification of arms control treaties. But when our space projects were being planned, they seemed as much like adventures as serious research. No doubt benefits from many of tomorrow's experiments will develop the same way.

One thing is very clear about future space research and travel: It will be even more expensive than in the past. It will be so expensive that few countries will be able to afford anything but small projects. Only the United States and Russia (the former Soviet Union) have their own major space programs. A group of countries in Europe has formed the European Space Agency (ESA) to share the cost of space

This is an artist's concept of a power conversion system. The structure collects pollution-free energy from the Sun and then transmits the energy to Earth in the form of microwaves.

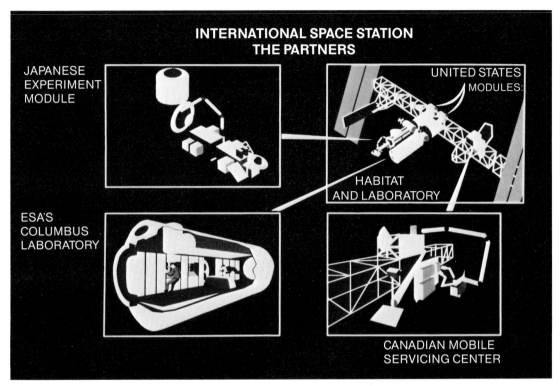

**INTERNATIONAL SPACE STATION
THE PARTNERS**

JAPANESE
EXPERIMENT
MODULE

UNITED STATES
MODULES:

HABITAT
AND LABORATORY

ESA'S
COLUMBUS
LABORATORY

CANADIAN MOBILE
SERVICING CENTER

Space station Freedom *is an example of different countries working together on a major space project. The international space station would consist of four modules and a servicing center.*

research. Canada and Japan also conduct limited space research. International cooperation is growing rapidly, and some of the biggest projects of the future may well turn out to be joint projects of two, three, or even of all five space programs.

Every new project is based on years of research. In many cases, new projects will require building bigger spacecraft that can carry people and equipment for experiments that take a long time to perform. In other cases, research will require completely new technology and brand new ideas. While that research is taking place, men and women will also be dreaming of projects for the middle of the 21st century and beyond. They will dream just as people did long ago when they watched and envied the birds.

The Aerospaceplane

Ever since the Wright brothers flew their first airplane at 20 miles per hour (32 km/h) in 1903, people have wanted to fly faster. Not only is speed thrilling, it is also a very good way to get places in a hurry. When traveling in the air, a vehicle needs both power and streamlining to move fast. In space, power is most important.

Our first taste of real speed came in 1947. In the rocket-powered Bell X-1 airplane, Chuck Yeager broke the "sound barrier" by traveling faster than the speed of sound, also called Mach 1. (The actual speed of sound depends on the altitude and the temperature of the air. At 40,000 feet [12,192 meters], for example, the speed of sound is normally 660 mph, or 1,062 km/h. Mach 1 would then equal 660 mph.) By 1967, W. J. Knight had flown Mach 6.7 (6.7 times the speed of sound) in the North American X-15. These flights were made at very high altitudes, where there was so little air that the airplanes behaved almost like spacecraft.

The first Concorde supersonic transport (SST) flew in 1969. Soon after that, it was used by commercial airlines for long-distance flights. The Concorde transports people across the Atlantic Ocean from Paris or London to New York or Washington at Mach 2. It made high-speed transportation available to travelers.

The next step in high-speed, long-distance travel—and it was a big step—was the space shuttle. This new vehicle is also called an *aerospacecraft*, since it is able to operate both in space and in the atmosphere. Launched by enormously powerful boosters into Earth orbit, it returns to land on a long runway as a powerless glider does.

The Concorde SST, top, *and the space shuttle* Discovery, bottom, *represent two important steps in high-speed travel.*

During the gliding part of its flight, it travels as fast as Mach 25, as it enters the atmosphere at an altitude of 50 miles (80 km). To protect the shuttle from the great heat caused by the friction of air against its outer surface, its most important areas are covered with special heat-resistant tiles.

While plans are being made to design and build larger and faster supersonic airliners, the real dreamers are hard at work on a long-range project that will make all other aircraft seem old-fashioned. It is the *aerospaceplane.* The United States calls it the National AeroSpace Plane (NASP), Great Britain calls it HOTOL, and Germany calls it the Sanger II. Whatever it's called, it's revolutionary.

The aerospaceplane will take off and land on a standard runway and fly more or less directly into orbit. The American and British aerospaceplanes will be SSTO (single stage to orbit), which means they will require no boosters. The German design requires a booster craft to launch.

There are several possible uses for this new type of Earth-to-space vehicle. It could replace the shuttle in ferrying astronauts and cargo to a space station in Earth orbit. It could become an orbiting spacecraft. It could be an

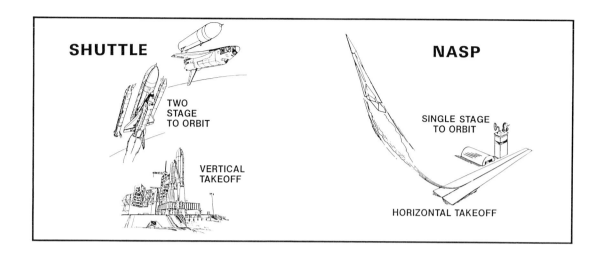

SHUTTLE

TWO
STAGE
TO ORBIT

VERTICAL
TAKEOFF

NASP

SINGLE STAGE
TO ORBIT

HORIZONTAL TAKEOFF

extremely fast airliner, capable of flying from New York to Tokyo in two hours at Mach 25.

Design of the Aerospaceplane

Generally, the current designs for aerospaceplanes look a lot like the Concorde SST, with a long, needle-like nose, sharply swept-back delta (triangular) wings, and engines tucked under the fuselage, or body. This shape works well at the very different speeds the craft must fly: a couple of hundred miles (or kilometers) per hour during takeoffs and landings and several thousand miles (or kilometers) per hour when cruising long distances.

The wings of the aerospaceplane would be steeply swept back to keep them inside the cone of the shock wave created when the plane's nose cuts through the air. The plane may even have a *blended* wing and fuselage. That means the flat-topped fuselage shape would merge with the wing. It would be hard to tell where one stopped and the other began.

The aerospaceplane would also need unusual engines. The Concorde SST uses fairly ordinary turbojet engines, much like those on any modern jet airliner. Turbojet engines don't use much fuel, but they can only work well up to about Mach 5. Rocket engines can operate at almost any speed in airless space, but they use tremendous amounts of fuel. An aerospaceplane has to fly at a very high speed and

The British HOTOL, top, *and the German Sanger II,* bottom, *are aerospaceplanes that can reach all low-Earth orbits.*

altitude, but if it used huge amounts of fuel, it would be too expensive to fly. Clearly, some new kind of engine is needed.

The SCRAMJET engine may solve the problem. It would be built right into the underside of the aerospaceplane and would use sections of the airplane as part of the engine. As the air passed under the front part of the fuselage, it would be slowed and then compressed. It would then enter the engine nacelle, or housing, where it would be mixed with fuel and exploded. The terribly hot exhaust would shoot out the back of the engine and be shaped for maximum thrust by the rear part of the fuselage bottom. If the SCRAMJET engine works as well as engineers hope, it could be the answer to a lot of problems connected with flying at very high speeds.

Operation of the Aerospaceplane

The aerospaceplane would work much like today's airliners—not like the space shuttle, which requires millions of dollars worth of special equipment and a crew of hundreds of people to prepare it for launch. The aerospaceplane would load up, taxi out to the runway, turn its engines up to full power, take off, and head directly for space. After hours or days at extreme altitudes, it would fly back to Earth and land on the same runway it had used for takeoff.

Although there would be scientific and military uses for them, aerospaceplanes could also be used as commercial airliners. An aerospaceplane used as an airliner would be called a *hypersonic transport*. It would cruise at hypersonic speed, which is over Mach 5. Such an airliner may cruise as fast as Mach 12 in the atmosphere and as fast as Mach 25 in orbit. It would be able to carry at least 200 passengers on very long flights, such as across the Pacific Ocean. It wouldn't make sense to use an aerospaceplane for short trips because it would have to spend most of the flight at low speeds and low altitudes, where it doesn't work efficiently.

Testing the Ideas

Before building a full-size aerospaceplane, ideas must be tested with research airplanes. The experimental X-30 is

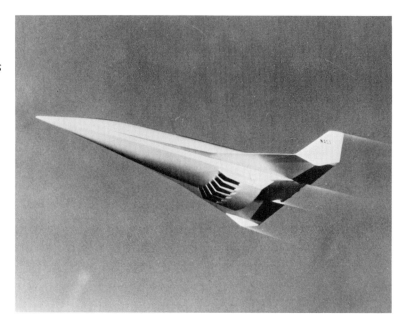

This underside view of an aerospaceplane shows the engine inlets. This design has winglets at its wingtips to improve efficiency.

In this artist's rendition, the National AeroSpace Plane is used as an airliner. One plane is taking off, one is landing, and another is being serviced.

being built to test the planned design for the aerospaceplane and the SCRAMJET engine. The engine would use hydrogen fuel instead of the usual gasoline or kerosene used in other airplanes. The X-30 is due to fly in the late 1990s.

If it flies the way the supercomputers used in its design predict it will, then the United States would probably go ahead and build the first full-size National AeroSpace Planes.

The U.S. lab is one of four modules in space station Freedom. *In the upper left, an astronaut is building extension booms up from the main boom, as part of Phase II.*

The Space Station

Building a space station is a project that would expand on existing research. The proposed space station would be a bigger version of an earlier project. It would carry more people and equipment and would be able to stay in space for a much longer period of time than earlier models.

The first temporary U.S. space station was *Skylab,* which was launched in 1973. Several crews of astronauts visited *Skylab*, conducted experiments, and learned to live in space for as long as 84 days.

However, the Soviets (in the former USSR) kept space stations up longer and learned even more. In 1971, the USSR launched the first of the *Salyut* space stations. They were larger and more elaborate versions of earlier piloted spacecraft, such as those the United States and the USSR had been orbiting

for several years. Using the new *Salyut* stations, the Soviets were able to keep their cosmonauts in space for as long as two months. They used a smaller craft called Soyuz to ferry supplies from Earth to the *Salyut* stations. The Soviets learned a great deal about the long-term effects of weightlessness (zero gravity) as they performed experiments and did mechanical work while adjusting to the strange conditions of working without gravity.

A few years later, a second generation of *Salyut* stations came into use. The new model had two docking ports to accommodate more supply craft. Having more supplies delivered made it possible to conduct much longer flights. But as more equipment, food, etc. was brought to the *Salyut* stations by Progress cargo craft, the stations became crowded and less pleasant for

On Skylab, *Owen Garriott performs a particle collection experiment during a space walk.*

the crew. A new and even larger space station was needed, and in 1986 the first *Mir* was launched. It is 108 feet (33 meters) long. With six docking ports, it can handle more supply craft as well as specialized research modules or units.

As space scientists learn more about living in space for long periods of time, they want even larger craft. Larger craft would be able to carry more people

who could perform complicated, long-term experiments to produce useful information. Many scientists believe that such a permanent space station can be expanded as needs arise. In the 1980s and early 1990s, scientists planned a space station that resembled a small village. This station was designed so that it could grow into a larger village with more people and more places to live and work. It could grow into a small town or perhaps a full-fledged city. This was to be the National Aeronautics and Space Administration's (NASA's) space station *Freedom*.

In 1993, the plans for this space station were changed because the U.S. Congress was not willing to pay the great cost of the original design. People in government started thinking seriously about building and operating a smaller space station in partnership with the Russian space program, as well as the space programs in Europe, Canada, and Japan.

The Russians already have experience with the smaller *Mir* space station that could be useful in this partnership. They also have large launch vehicles that could help lift pieces of the new station into orbit. After the huge political changes in their country in the early 1990s, the Russians had to cut back their own space program. This left them with many talented scientists and engineers with valuable contributions to make.

The new plans for the space station *Alpha* are on a smaller and simpler scale than the original designs for *Freedom*. Space station *Alpha* would not be able to carry as much scientific equipment or as many astronauts but would cost billions of dollars less. Much of what American scientists learned and planned for the *Freedom* project can be used in the joint U.S.-Russian space station and other space stations of the future.

Design of the Space Station Freedom

All previous space stations and crewed spacecraft have been boosted

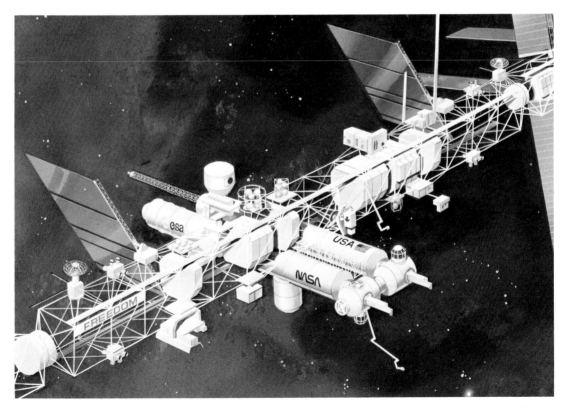

In space station Freedom, *four modules are attached to the main boom, two on each side. The Canadian servicing center is to the right of the module marked USA.*

A module is being assembled in space with the use of a free-flying craft to control the manipulator arm.

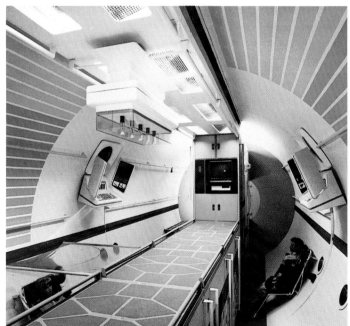

In this full-size mock-up of the interior of a space station Freedom module, an astronaut is strapped in to keep from floating in the microgravity environment.

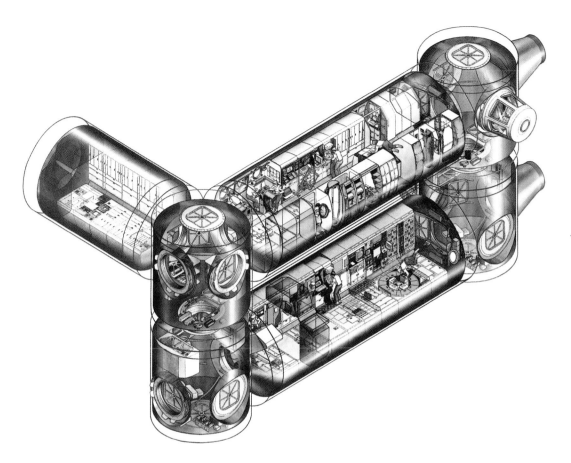

In this drawing, the habitation module is above the U.S. laboratory, with interconnecting nodes at both ends. Logistics storage is shown at the far left.

into orbit in one piece. In the original plans, the space station *Freedom* was to be built in space, piece by piece. It could begin operations before being completed, and it would grow as the need for larger facilities increased.

The basic Phase I of this space station consisted of three main parts: a horizontal boom, a cluster of modules, and solar panels.

The Horizontal Boom

The horizontal boom looked like a long section of bridge girder with many cross-pieces. The thin, light pieces of tubing used to build all its parts would be transported by shuttle flights as rolls of metal, which would be formed in space into tubes. The boom, about 500 feet (152 meters) long, was to be the foundation of space station *Freedom*.

The modules, the solar panels, and dozens of small experiment packages attached to the boom.

The Modules

At first, there were to be four modules, or units, representing the international nature of the station.

The *Habitat* is where the astronauts would eat, sleep, and bathe. It was to be designed and built for NASA, using knowledge gained from the U.S. *Skylab* and the European *Spacelab*.

The *Laboratory* would contain a working area where the U.S. astronauts could conduct experiments with equipment designed and built by government laboratories and private companies.

Columbus Laboratory was to be a joint project of all ESA nations: Austria, Belgium, Denmark, Germany, Ireland, Italy, the Netherlands, Norway, Spain, Switzerland, and the United Kingdom (England, Scotland, Wales, and Northern Ireland). This module was a fat cylinder 42 feet (12.8 meters) long and 13 feet (4 meters) in diameter. It would carry a payload of 22,000 pounds (10,000 kilograms). European crews would work in it after it had been carried up to the station as cargo of a space shuttle, in one piece.

The Japanese *Experiment Module* would be an unstaffed experiment unit. Astronauts wouldn't work inside it, but they would be able to service and repair it, remove completed experiments, and install new ones.

In addition to the four modules, a version of the National Research Council of Canada's mobile servicing center would be included. The servicing center would consist of a long arm which bends in the middle, something like your arm. It would be used for moving large pieces of equipment along the boom during construction and operation of the space station. A smaller Canadian-built arm has been used on the space shuttle.

Power for the Space Station

There would be two pairs of huge solar panels at either end of the long boom. These solar panels would collect sunlight that would be turned into electrical power for experiments and for the astronauts' life-support systems. Each panel would be 98 feet (30 meters) long and 33 feet (10 meters) wide. Altogether, the panels would have about 26,000 square feet (2,340 square meters) of solar cells to turn light into energy.

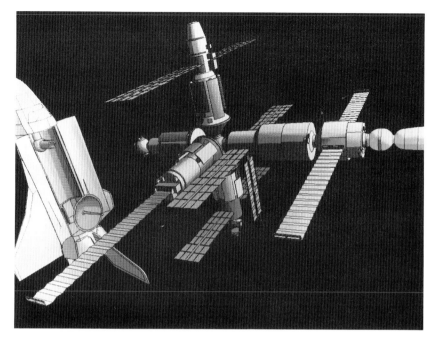

A 1993 sketch, left, shows one proposal for space station Alpha. *Here a shuttle is docking at a section of the space station that has been added to the Russian* Mir *station in Phase I of the international project.* Below left, Skylab's *huge solar panels are viewed from the command module.*

That's 20 times the area of *Skylab's* solar panels. The panels could be turned together or separately to keep them aimed at the Sun so they would generate maximum power.

Free-Flying Platforms

There would also be two smaller satellites used for space station experiments. These satellites would be sent into polar orbit (circling the Earth by passing over the North and South poles). One satellite would be developed by NASA and the other by ESA.

In Phase II of space station Freedom, *experimental and storage units would be built in a rectangle of booms.*

Functions of a Space Station

While the creation of a permanent space station would be a great challenge and a wonderful adventure, its main purpose is to make it possible to conduct experiments that cannot be carried out on Earth. Although the precise choice and purpose of future experiments done on a space station may depend on what we learn during its first months and years of operation, many areas of research have already been identified.

Biology

Since gravity plays an important role in the development of life, many space station experiments would be related to learning what happens to living things when there is little gravity (microgravity) or no gravity. Astronauts would try to learn if food can be grown on board future spacecraft during very long trips to Mars and other planets. They would try to learn if the absence of gravity causes serious damage to the human body. If it does, long flights could be dangerous to the crews or even impossible.

Astronauts would study the effects of microgravity on the cardiovascular system (the heart and blood vessels), and on muscles, bones, and on one's sense of balance. This study could lead to better medical treatment on Earth and to special medical facilities in future space stations.

Another goal of studying the long-term effects of microgravity on humans is to deal better with long flights in the future. Already, studies of low gravity on humans have led to better exercise programs for astronauts. If people are ever to journey to the planets, we must know the answers to many questions before the first trip is launched.

In the U.S. laboratory on space station Freedom, *the astronaut at the left is working in a commercial processing area. The two in the center are at a general work station, and a fourth astronaut is entering the module from a connecting tunnel.*

Physics and Chemistry

According to a NASA report:

The fields of physics and chemistry also benefit from the microgravity environment within the space station research laboratory. Again, gravity plays a fundamental role in many physical and chemical processes on Earth. An important example is the way gravity makes gases and liquids separate according to their density. In our weather patterns, we see the warm, thin air rise and the cool, dense air fall. On Earth, laboratory experiments are subject to the same phenomenon, and it is nearly impossible to force substances of different densities to mix perfectly.

In addition, the natural motion of the substance (called convection) can ruin experiments that require chemicals to remain perfectly stationary while slow reactions take place. However, in the microgravity of space, the problems of convection, of imperfect mixing, essentially disappear. We can produce near-perfect com-pounds in liquid, crystalline, and solid forms. Compounds of purity never before achieved will lead to new knowledge of the physical and chemical laws governing their formation.

Observing from Space

From the first time a human orbited the Earth, one of the most important tasks has been to observe the Earth and the atmosphere around it. The first astronauts and cosmonauts carried simple, hand-held cameras to record their rare view of the continents and the oceans, of clouds and storms, of forests and farmland.

Hundreds of Earth-orbiting satellites have been sent into space in the last 30 years, many of them to observe and report. As satellites have gotten larger, more complicated, and more sophisticated, they have been able to send back much useful information for scientists to study. Satellites, such as those that send back television or radar views of the weather, have become a regular part of our lives and can be seen every day on TV news programs. Photographs taken with a special camera, from a vantage point several hundred miles up, would record information that could never be obtained in any other way.

A large space station such as *Freedom* would be able to carry this wonderful work into a new era. The space station would have equipment for use in the sciences of astronomy, planetary studies, earth studies, plasma physics, and atmospheric physics. It would carry more and larger equipment than any previous spacecraft. It would carry that equipment longer and would have the ability to maintain and repair the equipment to keep it working at top efficiency.

Space station *Freedom* was designed to have thousands of square feet (hundreds of square meters) along its boom to attach observation instruments and the antennas needed to send data back to Earth. The usual orbiting spacecraft are limited in how much power they can generate to run their instruments and in how much information they can send back. But space station *Freedom* was to be very large, and it would have enormous solar panels to generate all the power needed to record information and send it back to ground-based laboratories. If more power would be needed, the "construction crew" of astronauts could simply add more or larger solar panels and antennas.

Space station *Freedom* could thus have been expanded, which would have given it thousands more square feet of structure.

Manufacturing in Space

Once astronauts have learned how to make better materials in space, it should be possible to begin limited production of some of them. Many types of materials might be produced, such as pure crystals, new high-strength metals, and temperature-resistant types of glass.

Purer crystals are important in the creation of smaller and more efficient communications equipment and computers. High-strength metals and temperature-resistant glass are needed for jet engines and optical communications systems, such as fiber-optic equipment for telephone communication. The separation of biological materials for making new types of drugs and medicines will also be possible in space.

Once production has been shown to be practical, it may be possible to build specialized manufacturing satellites. These satellites could be a part of the space station, but not directly attached to it. That way, valuable research space

In this drawing, large antennas are attached to the rectangular boom and to the main, horizontal boom.

on the station would not be taken up by commercial companies.

Space Station Support

The space station *Freedom* was not designed to be completely self-supporting. People who live in any community need deliveries of food (by truck to neighborhood stores) and water (by pipeline). They also need garbage removal. The astronauts would need the same kinds of services, except that their "trucks" would be space shuttles.

Materials from the laboratories in the space shuttle modules would have to be transferred to Earth laboratories, and waste products that can't be let loose in space would have to be removed. Astronauts whose tours of duty have ended would have to be taken back to Earth, and new astronauts would be sent to replace them. These things would be done with a routine shuttle service using both the well-known NASA space shuttle and later an improved shuttle that could carry a bigger load. Also, ESA is developing its own shuttle, to be called *Hermes,* for resupplying the space station. There could also be some one-way flights of multistage cargo rockets to deliver needed products.

Life aboard a Space Station

Men and women will be living aboard a space station for long periods of time: working, eating, sleeping, and doing leisure-time activities. If they are to work at top efficiency, they must remain healthy, happy, and on reasonably good terms with each other. If they become unhappy because they find the food unsatisfying, or the working conditions become difficult, or they have trouble sleeping, or they are just bored, then the experiments they are working on may not be performed correctly. The entire project could fail if the crew's morale fails.

Eating is an especially important part of life not only because it provides the fuel for living, but because it is such an enjoyable, familiar experience. In the early days of space flight, astronauts put up with some pretty dull (and sometimes unpleasant) food and drink. But since their flights were short and very exciting, they accepted poor-tasting food as part of the job.

On long flights, however, food and drink will have a great impact on the crew's morale and health. In space, all food must be specially prepared, somewhat like airline food, but it must taste a lot better. Airline passengers rarely have to eat more than one meal on a flight, but the space station crew will have to eat on board for weeks or months at a time. The food must be not only nourishing, but also tasty, varied, and interesting. On international space flights, each astronaut may be allowed to bring along some foods common to his or her culture. A Russian cosmonaut, for example, might want to take *borscht*, a soup made from beets. A typical day's menu for one member of a recent U.S. shuttle flight included:

Breakfast: beef patty, bran flakes, breakfast roll, peach yogurt, and decaffeinated coffee.

Lunch: shrimp cocktail, crunchy peanut butter, grape jelly, tortillas, trail mix, and tea with lemon.

Supper: sweet and sour chicken, rice pilaf, asparagus, butterscotch pudding, and tea.

In addition, the pantry was well stocked with many snacks, such as fruit drinks, nuts, cookies, candy, pudding, fresh fruit, crackers, and granola bars. That was for a seven-day mission.

Russian cosmonauts might enjoy chicken with prunes, honey cake, sauerkraut soup, and cottage cheese with pureed black currants.

On a long trip, to Mars for example, stocking fresh fruit would be out of the question. It would be necessary to grow some food in the spacecraft because there wouldn't be room to store food to feed several people for many months. Among the foods that have been successfully grown on a spacecraft on an experimental basis are vegetables such as cucumbers, tomatoes, peppers, peas, parsley, and lettuce; fruits such as strawberries; and other foods such as garlic, oats, and herbs.

Much research remains to be done on growing plants on a long mission. Food weight and nutritional value would have to be considered and balanced with appeal. Whoever plans the meals would have to be very clever to keep them from becoming routine and boring.

Sleeping aboard the space station would not be a simple matter of finding some quiet place to stretch out and then turning off the lights. There is no "down" in space. Crew members would be able to sleep in any position, but they would have to tie a sleeping bag to a wall to keep it from floating away. Much has been learned about sleeping in space from long flights in United States and Soviet spacecraft, so this should pose few problems.

Personal Hygiene

The problems created by poor personal hygiene during a long space voyage are much more serious than on a short orbital flight. Little things can quickly grow into major problems if they aren't taken care of quickly. In this area, the Soviets learned a great deal on missions lasting several months.

Washing is a part of life that is second nature on Earth but very different in space. Water refuses to flow in zero gravity. It forms small droplets that float around in the spacecraft and can easily work their way into delicate electrical devices. If they aren't controlled carefully, the droplets can cause short circuits. Neither a washbasin nor a bathtub will fill up, because there isn't enough gravity to hold the water down. Water has to be sprayed on an astronaut inside a container and then removed with a small vacuum cleaner. However, on long space flights frequent showers would use more water than can be stored on a spacecraft.

Soviet experience suggests that an astronaut can keep clean by taking a shower once every 10 days. In between showers, the astronaut can use towels made of antibacterial cloth soaked in a disinfectant. But unless this washing system works very well, it could make some crew members very uncomfortable and lead to a breakdown in teamwork.

To wash their hair, astronauts must use a towel attached to a massage brush, though the results are not quite like using an Earthbound shampoo. People who are unusually sensitive about cleanliness may have to be weeded out of programs involving long stays in space.

Brushing teeth is easier with a battery-powered toothbrush and nonfoaming toothpaste. Using chewing gum after every meal also helps keep teeth clean.

For shaving, a normal electric razor with rotating heads works very well, as long as it has an added nozzle to vacuum the shaved whiskers. Otherwise the cut whiskers would fly around the inside of the spacecraft and find their way into tiny openings in the equipment, where they could interfere with electrical connections.

Something as simple as the way the air in the spacecraft smells can make the difference between a successful long space mission and one that fails. The Soviets discovered that certain smells, like garlic, can be very difficult to remove from the cabin air, even with their best air-conditioning equipment. If the air would smell for several days after garlic is used in a meal, it prob-

ably would be wise to eliminate the garlic from the food, even if it means the food would taste flat.

The science fiction image of all the members of a spacecraft crew dressed in neat, identical uniforms like the crew of a naval vessel does not fit reality. Soviet experience suggests this is an area where the individual's taste can play an important role in keeping morale high. The Soviets recommended that each crew member be allowed to pick clothing colors and even to take part in designing the styles, location of pockets, and decorations of the clothes.

Washing clothes poses a major problem. In space it would require a large amount of water and special facilities for washing, rinsing, and drying clothes. So far, long-term missions have used throw-away clothes that are worn for a specified number of days and then packed away until the mission ends. This is one field in which there is a lot of opportunity for new ideas.

Exercise in space is similar to exercise on Earth, but is even more important. Normal activity doesn't strain the muscles in space because nothing has weight, and lifting a heavy object takes no more energy than lifting a piece of paper. Muscles that aren't used get

Pilot William Pogue is holding on while Commander Gerald Carr passes trash bags through an airlock aboard Skylab, *left. Below, one astronaut is strapped down so he can sit while eating his lunch. The other removes her meal from a microwave oven.*

The mock-up habitation module has sleeping quarters with sliding doors for privacy. The sleeping bag, far left, *is attached to the wall to prevent the astronauts from floating.*

weak, and bones lose their strength. That means a regular program of exercising is vital. During the *Skylab* mission, the astronauts exercised at least 30 minutes each day with stationary bicycles or with springs (rather than weights).

Recreation can mean the difference between a happy, smoothly working crew and a crew that may become bored, or even bitter. Each individual needs his or her own choice of music, books, videotapes, and other familiar ways of spending spare time. Those who enjoy going for a run through the woods or for a swim will be out of luck. But sightseeing is outstanding from space. The view from the windows is one of the most wonderful experiences imaginable.

The Lunar Base

The Moon has been watched, worshipped, and studied since ancient times. Interest in the Moon must have begun the very first time someone looked up and realized the Moon was different from the stars and planets. The viewer could see details, not just a light. In ancient times, people didn't know what the Moon was, or how far away it was, or how big it was. Its position shifted and its shape seemed to change, but the Moon's cycle was always repeated. Astronomy, one of the oldest sciences, began when Galileo used a crude telescope for the first scientific study of the Moon's surface in 1609. We've found out much more about the Moon since then.

In the late 1940s, the U.S. Army Signal Corps "made contact" with the Moon by bouncing radar waves off its surface. The Soviet Union launched *Luna 2*, the first spaceship to reach the Moon, in 1959. *Luna 3* sent the first pictures of the Moon's surface back to Earth. Little by little, we learned more about the Earth's natural satellite, only 238,000 miles (382,942 km) away.

In 1962 President John Kennedy announced Project Apollo, aimed at landing astronauts on the Moon and bringing them home by 1970. The United States launched a series of Ranger space probes that sent back the first close-up pictures of the Moon's surface. From 1966 to 1968, the United States landed five Surveyor spacecraft on the Moon. These lunar probes took many detailed photographs and sent back information on the Moon's composition.

At about the same time, *Apollo 8* astronauts made several flights around the Moon to search out likely landing

Edwin (Buzz) Aldrin sets up passive seismic experiments to detect moonquakes after the Apollo crew leaves the Moon.

sites and learn still more about the conditions on the surface. Boosted aloft by the huge Saturn V launch vehicle, the *Apollo 11* consisted of a cone-shaped command module, a cylindrical service module, and the spindly Moon lander. On July 20, 1969, Neil Armstrong and Buzz Aldrin stepped out of the Moon lander and walked on the Moon. They were the first humans to visit another world.

More Apollo flights produced an enormous amount of information about the Moon, including hundreds of pounds of Moon rocks and dust that are still being studied in laboratories.

In the process, we learned how to survive and to work on a planetlike body that has no air, no water, and very little gravity (one-sixth as much as the Earth). Unfortunately, the cost of the Apollo program was so great that it ended without any plans to continue exploring the Moon after the final Apollo trip in 1973.

Back to the Moon

In the late 1980s, serious talk began about returning to the Moon for long-term studies and perhaps even building a permanent base there. The Moon is an ideal place to learn more about living on other planets since it resembles them in many ways, and it is so close to Earth. Its lack of atmosphere would make it a great place from which to observe distant objects, because there would be no interference from smog or from city lights as there is on Earth. And the reduced gravity would make construction of living quarters or laboratories much easier, since everything would be lightweight—about one-sixth of what it is on Earth—including construction material and equipment.

Two plans are being considered for the Moon in the early part of the 21st century: a lunar observatory and a permanent base that could be used for conducting experiments and for launching or refueling flights to Mars and other distant planets and stars.

The Lunar Observer

Before any steps can be taken to establish an observatory or a base, we would have to learn more about the Moon. Sometime toward the end of the 1990s, a lunar geoscience observer may be put into orbit around the Moon. It

This illustration shows a fixed radio telescope built into a crater. To the right are movable radio telescopes. An astronaut is walking toward a small optical telescope.

would map the Moon's surface and study the makeup of the soil. Depending on what the lunar observer finds, it might be followed by uncrewed landers and remote-control roving devices. The goal would be to find places to set up an observatory and a base.

Design of the Lunar Observatory

The observatory would probably be built on the far side of the Moon, where instruments would not be bothered by interference from the Earth's light and radio emissions. Construction materials and workers would be ferried to the Moon base via the space station. The workers would land on the Moon in special vehicles, which they could use as living quarters while they worked.

Research facilities would include equipment to detect, measure, and record lunar activity, solar activity, and deep-space activity. Building all this would require specially designed machinery to move large amounts of soil. Astronauts would also need one or more rover vehicles—larger versions of the lunar rovers driven by some of the Apollo astronauts.

Because of conditions on the Moon, the construction work would have to be limited to the lunar day, which lasts 14 Earth days—almost half a month. The rest of the time the Moon is so dark that construction work would be almost impossible.

Operation of the Lunar Observatory

The observatory would be occupied by astronaut-scientists who would spend short periods of time there. A typical mission would last no more than 20 days. Six days would be spent traveling to and from the Moon (three days each way), and about 14 days on the Moon would be spent working.

The astronauts would live in the lander vehicle. They would move around in open, four-wheeled buggies able to take them as far as six miles (10 km). They would check and, if necessary, repair the permanent instruments, explore the surrounding area, and gather samples.

Types of Lunar Research

Optical Astronomy. Conventional telescopes that use lenses would work much better on the Moon than on Earth. That is because, due to the lack of atmosphere on the Moon, the air is

This huge array of small radio telescopes works the same as one very large telescope.

not disturbed. Large telescopes—such as the Hubble Space Telescope—would be put into Earth orbit. But even larger ones, built on the Moon, would work best of all. They could be put into place as the lunar observatory is being built, and their information could be radioed back to Earth or to the space station.

Radio Astronomy. Radio astronomy uses very large antennas, some of which are on rotating stands. Other antennas are built right into the ground. They, too, could be built bigger and would work better on the Moon because of the lack of interference. Information

from the antennas would be sent back by radio, and the antennas would probably need only periodic maintenance.

Geology. The study of the Moon's solid matter is called *selenology,* but it is almost exactly like geology, the study of the Earth. Instruments could be placed miles from the main part of the observatory, to transmit their information back for analysis on Earth.

Life Sciences. While the astronauts were making their regular visits to the lunar observatory, they would be conducting tests on themselves to see how they performed in the alien environment. Knowing how these lunar

The lunar lander, right, is leaving an orbital transfer vehicle that brought materials from the space station to the lunar orbit. The lander will take some materials to the Moon's surface and then return for more.

astronauts survive in an environment without air or water would help those who are planning longer stays on the Moon and short stays on Mars and other planets.

A Permanent Lunar Base

A permanent lunar base would be a much more ambitious project than an unstaffed observatory that would be visited only for short periods of time. A permanent base would depend on what astronauts learned at the unstaffed base and its surrounding land. A permanent lunar base would serve several purposes: lunar studies, human studies,

and a base for future flights to Mars and elsewhere in space.

A permanent base would make it possible to use the Moon as more than merely a place to experiment and explore. The Moon would become a place to do things—an extension of the Earth.

Design of the Permanent Base

The first part of the lunar base to be built would be a domed building, used as living quarters for the regular crew and for any visiting astronauts. It would be fully pressurized (like an airplane), climate-controlled (for protection against the extreme temperatures), and as comfortable and like the astronauts' Earth homes as possible. Since the Moon has some gravity, the crew would be able to sleep in beds, eat at a table, and take showers.

Little by little, a science center would be created, starting with fairly simple experimental equipment and progressing to larger and more complicated units. Some equipment would be near the habitat, or living quarters, so it could be serviced easily. Other equipment would have to be miles away because of the terrain or because some of the equipment would have to be free of interference from other equipment.

At this landing facility, the pressurized vehicle at the lower right is connected to the lunar lander by a flexible tunnel. The astronaut at the bottom is preparing to change an engine. The solar-powered crane at the left is used to remove equipment from the lander.

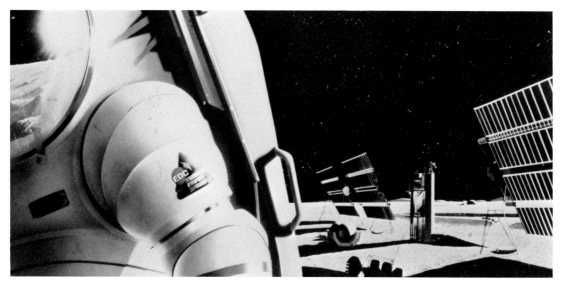

In this illustration, the astronaut to the left, who works for a commercial firm, is checking solar equipment.

Operating and maintaining all of the equipment would be one of the most important duties of the permanent crew.

The Moon's mineral resources, the actual material out of which it is made, could also be put to use. After scientists learn what lunar soil and rocks are made of, astronauts could set up the first extraterrestrial (away from Earth) industry. They would be able to organize the mining of lunar materials, process these materials, and then store them until needed.

Much of the construction on the Moon would be for power-generating equipment for the lunar base. This equipment would consist of large solar power units as well as nuclear power units shipped from Earth in complete or nearly complete packages. As the base expands, its need for power would increase.

Lunar Construction

To build a permanent base, more material and equipment would have to be transported to the Moon than has ever before been sent into space. Such a large project would require a completely new type of launch vehicle, a cargo craft that is propelled by nuclear-electric power. That type of power would be much more efficient than the solid- and liquid-fuel rockets that have boosted the space shuttle, *Skylab*, and other spacecraft.

A series of flights carrying people and cargo would have to be made on a regular schedule to build up the lunar base. They would probably use a new low Earth-orbit transfer facility—a special kind of space station just for transferring fuel and cargo to Moonbound craft.

Operation of the Lunar Base

The base would use local materials—it would "live off the land"—as much as possible. For example, if oxygen could be extracted from lunar rocks, it could meet the needs of the crew's life-support equipment and also be used for rocket fuel. Using local materials would reduce, or even eliminate, the need to transport large quantities of oxygen from Earth to the Moon. Launch vehicles would only be used to transport materials not found on the Moon.

Missions to Mars

The planet Mars is the glamour target in space. For many years, Mars was the place where fiction writers based their imaginary alien characters. The "Red Planet," as Mars is sometimes called, is still wrapped in mystery and romance that will continue until astronauts are able to explore its surface and report their findings. The romance of our neighbor planet may not even end then, if explorers find signs of past or present life on what now appears to be a dead planet.

Mars is more like Earth than any other planet in our Solar System. Its day is just 37 minutes longer than Earth's, its gravity is almost 40% of that on Earth. Mars has a great deal of water trapped in its frozen soil and in minerals in the ground. But there is almost no oxygen in the Martian atmosphere, and the air pressure is very low. But Mars still wouldn't be as unpleasant a place as the Moon for future astronauts to work and live. Mars has seasons, clouds, volcanoes, and sand dunes much like Earth.

Past Missions to Mars

Scientists have been sending uncrewed spacecraft to Mars since 1965, when the *Mariner IV* flew as close as 6,118 miles (9,844 km) to Mars and sent back the first information. By 1969, *Mariners VI* and *VII* had flown within 2,500 miles (4,023 km) of the Martian surface and returned more than 100 pictures with more detail than had ever been seen before.

The Soviet Union landed its *Mars-2* craft on the planet in 1971 and *Mars-5* in 1975 but these missions didn't gain much information. In 1976, however, America's *Viking 1* and *Viking 2* landed on Mars and sent back color photos of the rocks and soil. The photos gave us our first close-up views of another planet. The Viking spacecraft also scraped up the soil and analyzed it in a remote-control laboratory aboard the landed craft. Tests showed no solid evidence of life on the planet, though the results were not completely clear.

So much was learned from the information gathered from the spacecraft that have orbited and landed on Mars that scientists concluded it would be safe for astronauts to land there. The surface of the planet is solid rather than deep dust, and there was no reason to expect any kind of danger to explorers. Plans went ahead for future crewed missions to Mars.

The search for information continues. In 1993, NASA's Mars Observer was about to orbit the planet when ground controllers lost contact with the probe. No one knows what happened because it sent back no information. This has delayed plans to send more advanced probes and astronauts to Mars. Scientists need more years of research before astronauts are sent there.

Flight to Phobos

Mars has two very small satellites, or moons, named Phobos and Deimos.

These moons look like good places for an advance group to explore before journeying to the Martian surface. Phobos is the current top choice. Although its surface has a lot of craters, as our Moon does, Phobos is much smaller (13 by 17 miles, or 21 by 27 km). It orbits a mere 3,700 miles (5,953 km) above the surface of Mars.

What makes Phobos so interesting to scientists is its soil, which appears to contain a lot of water. Even more important, the soil also appears to contain carbon-based materials that could be made into rocket fuel to be used for later flights to Mars. The low gravity on Phobos would make landings and takeoffs relatively easy. The moon's nearness to Mars would make it an ideal base for studying that planet, and Phobos would be an excellent stopping place on the way to Mars. Also, a crewed mission to Phobos would be good practice for a trip to Mars.

The Trip to Phobos

Going to Phobos would require two very large, powerful spacecraft. The first would be an uncrewed cargo hauler. It would carry the exploration equipment, the remote-control rover vehicles, and the fuel for the return flight

Above, *this photo of Mars was taken looking south from* Viking 2, *which can be seen in the lower left corner. The largest rocks are three feet (1 meter) across. Many are porous or spongy.* Below, astronauts explore Phobos in personal vehicles (left and lower right). *The lander is at the lower center, and the Mars orbiter can be seen above it.*

A Mars rover vehicle digs into the surface and sends data and photos back to an orbiting craft. From the orbiter, the information is sent to Earth.

of the crewed spacecraft. The second craft, which would be carrying four astronauts, would leave Earth only after the cargo craft was safely in orbit around Mars. When the piloted craft reached Mars, the two crafts would link up in Mars orbit.

Two of the astronauts would use a small, specially equipped craft to reach the surface of Phobos, where they would remain for about three weeks. The other two astronauts would stay in Mars orbit and work from their spacecraft.

Phobos Exploration

Although the almost total lack of gravity on Phobos would make landings and takeoffs easy, it could also present a problem to exploring astronauts. A small car that weighs 2,500 pounds (1,136 kg) on Earth would weigh only 2.5 pounds (1 kg) on Phobos—about as much as a quart of milk weighs on Earth. With so little gravity, astronauts would land softly, and climbing and jumping over large rocks or wide craters would be easy. But a low level of gravity also means astronauts would have to tie down their craft to keep it from being knocked over by a casual nudge.

Once the first two astronauts arrived on Phobos and converted their lander into a safe home, they would be ready to explore that strange little world. The astronauts would use a new kind of personal vehicle, a cross between a piloted, maneuvering unit (the type used for space walks) and a moon buggy, an open buggy that bends in the middle (like the lunar rover used by the Apollo astronauts). Each astronaut would wear a full-pressure suit—an inflatable suit to protect the body from low atmospheric pressure during space flight—with an attached life-support system. A set of small rocket motors would help the astronauts move around the airless space of Phobos.

While on the tiny moon, the two astronauts should be able to explore most of the satellite, conduct experiments, and collect rock and soil samples for laboratory tests. By the end of their mission, they should have a pretty good idea of the nature of Phobos and what is involved in exploring a world with almost no gravity.

While those astronauts were busy on the surface, the other two astronauts could explore Mars from a distance. The orbiting astronauts could send a pair of remote-control rover vehicles down to the surface of Mars. Since the astronauts would be within a few hundred miles (or kilometers) of Mars, they could control the rovers directly, or in "real time." If the rovers were controlled from Earth, it would take at least seven minutes for a TV picture to travel from Mars to Earth to show what a rover was doing and another seven minutes for a signal from Earth to reach the rover. By the time the Earth signal reached the vehicle, it could have driven off a cliff and been destroyed.

Information from the Mars rovers would be sent back to the orbiting spacecraft in the form of data and pictures to be recorded for later study. Samples collected by the rovers would be returned to the spacecraft at the end of the Mars surface mission.

Return to Earth

Once the astronauts had completed their work on Phobos, they would return to the Mars orbit and prepare for the six-month trip back to Earth. After transferring the fuel from the orbiting cargo craft to the piloted spacecraft, the spacecraft would blast out of Mars orbit for the journey of more than 50 million miles (80 million km) back

A Mars lander is about to touch down.

to the space station in Earth orbit. There, the astronauts would switch to a space shuttle and complete the trip to Earth.

The Crewed Mission to Mars

After all the information gathered from the Phobos trips has been studied, a crewed mission to Mars might be planned. Due to the enormous cost of such a mission, it probably would be necessary for the United States, Russia, and the ESA to cooperate in a joint venture. A small step in that direction was taken in the Soyuz/Apollo mission of 1975. For that mission, the U.S. Apollo command module docked in orbit with a Soviet Soyuz spacecraft, and astronauts flew with cosmonauts for several orbits. No other joint Russian/American flights have been made, but both countries have been looking seriously at cooperation for future long-distance space missions.

It is very possible that major advances will be made in space propulsion that would cut the cost and/or the time of a Mars mission so much that one country could afford to do it alone. Other countries might then add their experiments and an astronaut to the mission.

Getting to Mars

Although it would be complicated and expensive, NASA is thinking about a series of three complete Martian trips, several years apart. As with the Phobos mission, each trip to Mars would involve two large spacecraft, one uncrewed cargo craft and one crewed spacecraft. The cargo ship would carry fuel for the return voyage and a landing vehicle, which would include a habitat or living enclosure, exploration equipment, and an ascent vehicle.

Because of the great loads to be carried, the two Marsbound spacecraft would probably need to use a transfer facility. In orbit like the space station, the transfer facility would be used for assembling both the cargo and crewed craft and for fueling them for the long trip.

Astronauts on Mars are laying out a solar panel to provide power for experiments and for their lander/living quarters, left.

The cargo ship would be sent on the eight-month journey to Mars first. When it arrived there, it would be placed into Mars orbit. After the cargo craft was safely in orbit, the crewed craft would be launched from the transfer facility. Eight men and women would travel for about seven months from Earth orbit to Mars orbit. In Mars orbit, they would meet and link up with the cargo craft that would be waiting for them.

Four of the astronauts would then descend in a landing vehicle onto the surface of Mars. The other four astronauts would stay in the spacecraft where they could conduct experiments and monitor and assist the four astronauts on the surface of Mars.

At Work on Mars

The astronauts on the surface of Mars would have about a month to learn as much as possible about the planet. They would live in a specially designed, prefabricated building, one whose parts had been made at a factory. The astronauts could erect the building quickly, since its construction would consist mainly of assembling the standardized parts. After they had explored the surrounding area, they could grad-

ually move farther away on foot or in a Mars rover vehicle. The astronauts would set up equipment that would operate automatically for many years after they had left, sending back data on the changing Martian conditions. Such equipment was tried on the Moon and has worked very well.

The astronauts would make detailed maps of Mars, and they would also conduct studies in geology, biology, and climatology—scientific fields they will have studied for many years before going to Mars.

Geology. Geology is a science that deals with the history and life of an area as recorded in rocks. Geologists also study the features (mountains, plains, etc.) of an area. The astronauts would dig into the soil on Mars—into the hills, the valleys, and the canyons—in an effort to learn the secrets of millions of years of Martian evolution.

Biology. Mars is the most likely planet in the Solar System to have living organisms. There may not be anything living there today, but there may have been life thousands or even millions of years ago. Scientist-astronauts would search for fossils of ancient life as well as for any clues to primitive life below the surface, where there may still be enough water to keep tiny organisms

This Mars robot vehicle, left, is powered by a nuclear generator. The vehicle is equipped with stereo cameras and devices to detect possible danger to itself. The data it collects would be sent directly back to Earth. In the simulation below, two scientist-astronauts have found ice at the bottom of a drill bit that had been sunk into a frozen crater near the south pole.

The ascent vehicle is blasting off from the lander/base to return to Mars orbit, where it will dock with the orbiting craft.

alive. Finding a few bacteria would be the first proof that life exists beyond the Earth.

Climatology. The Sun heats the northern and southern halves of Mars unequally, causing changes in seasons and temperatures. Surface temperatures on Mars range from −240°F (−151°C) during the Martian winter to 75° F. (23.8°C) at noon on a hot day at the equator during the Martian summer. Polar ice caps, which appear white from the Earth, are located at the north and south poles of Mars. The planet also has dry, desertlike regions that are covered with dust, sand, and rocks. Astronauts might establish a small weather station to record the speed and direction of the wind, the changes in temperature, the amount of sunlight that falls, and the ways clouds develop, move, and change.

Resources. Just as the Moon may be mined for materials that could be used in the lunar base and elsewhere, Mars may also become a source of raw materials. If enough resources are found, it may be possible someday to have a permanent base on Mars that would need very little support from Earth.

Going Home

After studying Mars for a month, the four astronauts on its surface would return in the ascent vehicle to the orbiting spacecraft. Their other equipment would be left behind for the next team of explorers.

After powering out of Mars orbit, the astronauts would begin the six-month trip back to the vicinity of Earth, where they would dock at either the space station or the transfer facility. There, they would unload their data, photographs, and videotapes and fly back to Earth in a shuttle craft.

A space shuttle docks at space station Freedom. *Part of a solar panel can be seen at the left.*

The 1,000-foot (305-meter) dish antenna that has been built into a natural bowl near Arecibo, Puerto Rico, searches for intelligible signals.

The Search for Extraterrestrial Intelligence

All of the ideas considered in this book so far have dealt with humans and our future plans in space. But for as long as anyone knows, we have been looking at the sky and wondering if anyone was "out there." Almost every planet (including those in and those outside our Solar System) holds the possibility of life—even intelligent life. Since the Earth is teeming with life even in its most desolate areas—deserts, high mountains, polar regions, the bottom of the oceans—why, we wonder, can't there be life out in space? And if there is life, can that life be intelligent and able to communicate?

Until scientists developed the equipment and the techniques for studying the Moon, planets, and stars, anyone could claim to know all about Moon people or Martians or Venusians or star dwellers. Who could say that they were wrong without having some basis in fact?

Study of the nearest planets—Venus and Mars—provides some hints. Because Venus is so much closer to the Sun than is the Earth, it is terribly hot. Also, the atmosphere of Venus creates a greenhouse effect, trapping heat near the surface of the planet. Mars, on the other hand, is more like Earth, with reasonable gravity, some water, and temperatures that are within reason. A lot more information about Mars will be needed before anyone can say there is no chance of any kind of life there. As for the more distant planets, even less is known. All of them are so far from the Sun that they must be much too cold for humanlike life. Some of them are extremely large and have very strong gravity. But some of them have moons that are similar in size to the Earth, and some of those moons just might support some kind of life.

Beyond the Solar System, there are millions of stars (like our Sun), and any star could have planets circling around it. But until one of these planets has been seen, their existence is just an idea. There could be a lot of them, and some could be very much like Earth.

Over the past 30 years, a new scientific project was developed, called SETI, the Search for Extraterrestrial Intelligence. (SETI was renamed the High Resolution Microwave Survey in 1993.) Its purpose was to find out if there is anyone or anything "out there." Trying to communicate with extraterrestrial (originating or existing outside the Earth or its atmosphere) life is a very difficult task. Because no one knows what extraterrestrial life is like, it's hard to even imagine starting a conversation.

Why Try to Communicate?

If we find intelligent beings, will it be possible to ignore them? Could we just sit back and pretend they weren't there? Or would we have to try to communicate?

If we find an intelligent being out there, we'll probably want to exchange greetings and see where that might lead. What amazing things could be learned in one brief exchange with a being living many millions of miles away!

There could be so much to learn: Do they have any form of society? If so, what type? What about their science and technology? Do they produce any of what we call the arts? Do they have the equivalent of music, literature, or painting? If not, do they expend energy on things we can't even imagine? What is their history? How do they move from place to place? Learning the answers to such questions could benefit humans enormously. It could lead to better technology as well as to new ways of looking at ourselves and our problems. Perhaps we would find ways to think about things that we can't even imagine now.

Is it also possible that learning all these things could be a big mistake? What would happen if these other beings are so far advanced that they could quickly solve all our problems, leaving us nothing to work on? What would our smartest, most creative men and women do? What would be the point of getting an education? Or of trying to make the Earth a better place? Another danger is the possibility that the extraterrestrial beings might be

This 210-foot (64-meter) parabolic dish antenna is one of several at Goldstone, California.

aggressive and warlike. What then?

Until we make contact with another civilization, we will have no idea how communications with extraterrestrials will work out. In the meantime, a lot of people are trying to find out if there *is* anyone out there. Some researchers are listening for signals from other worlds, while others are searching for other clues to life beyond the Earth.

Searching for Clues

If there is any life elsewhere in space, maybe some clues can be detected near the Earth as well as out in deep space. If we could find any solid evidence of life elsewhere, then it would be a lot easier to justify a much greater search.

Meteorites are pieces of material that are found on Earth but that come from outer space. They are the part of a meteor that isn't completely burned during the fiery plunge through the Earth's atmosphere. A meteor is seen at night as a bright streak of light in the sky, or a shooting star. Most science and natural history museums have at least one meteorite.

Scientists in laboratories split meteorites to look for anything that might indicate some hint of life. Sometimes they have found organic materials— what they think might be signs of life —deep inside meteorites. But they know that there is always the chance that some Earth material might have gotten into the meteorite when it entered the Earth's atmosphere, while it was on the ground, or even while it was in the laboratory.

Of much more scientific importance is the detection of some 150 types of organic molecules in deep space. Organic molecules, which are called the "building blocks of life," are groups of atoms that include at least one carbon atom. They aren't living things, to be sure, but they are close. When some organic molecules combine with others, they can form amino acids, which are found in or produced by organisms that can reproduce themselves, such as living cells. Astronomers have picked up sure signs of organic molecules in space.

Just because the building blocks of life are apparently in space in great quantities doesn't mean these blocks have been built on each other to form living things. But the strong possibility of life exists, and it apparently exists throughout space.

To find out if complex molecules can produce life, scientists on Earth have built simulations of planets in large, sealed jars. A science called *exobiology* has developed, as more and more tests have shown that molecules known to be in space can resemble living material if the conditions are right. Scientists put Martianlike soil into a jar, add gases to make a Martian atmosphere, and then inject some organic molecules. When all this mixes together, some forerunners of life can be detected.

Searching for Signals

If very basic signs of life developed in space a long time ago, then maybe the life cycle has evolved beyond complex molecules into small animals, large animals, or even intelligent beings who may be looking for us while we're looking for them.

There is certainly no lack of places to look. Scientists estimate there could be as many as 300,000,000,000 stars just in our Milky Way galaxy. Many of those stars could have planets, and some of those planets could be enough like Earth to permit advanced life to develop. And think about this: There could be as many as 10,000,000,000 other galaxies! Altogether, that means there could be 3,000,000,000,000,000,000,000,000 (3 sextillion or 3 million trillion) stars that just might have planets around them where life may exist!

The real problem is distance. The nearest star system that might have planets is 4.3 light-years away. (One light-year is the distance light travels in a year at 186,000 miles, or 299,274 kilometers, per second.) That means it would take 4.3 years for a radio signal to reach Alpha Centauri and another 4.3 years for an answer to come back. And that's the closest star. The others are so much farther away that signals from many of them would take millions of years to make the round trip.

Since sending signals to other planets is not a very likely way to learn of other civilizations, scientists have concentrated on another idea. They have been listening, in hopes of hearing someone else's signals. They use radio telescopes, which are really giant receiving antennas. These telescopes are highly sensitive, even to very weak signals coming from great distances.

The First Searches

In 1959, Project Ozma (named for a princess in L. Frank Baum's Oz books) began at Green Bank, West Virginia, under the direction of Dr. Frank Drake. For four weeks he used an 85-foot (26-meter), dish-shaped receiver to listen for signals from two of the nearest stars: Tau Ceti and Epsilon Eridani, both just 12 light-years away. Nothing was heard, but it was the first step of a very long search.

The Soviets began a similar study at an observatory near the city of Gorky in 1968. Receivers there listened for signals from 12 stars. Over the next few years other searches were also made. A 1,000-foot (305-meter) antenna was built into a crater at Arecibo, Puerto

The 230-foot (70-meter) receiving dish of this radio telescope, part of the Deep Space Network, is located in Canberra, Australia. The radio telescope is used for southern hemisphere coverage during microwave observing projects. The rotating part of the dish weighs more than 6 million pounds.

Rico. In 1975 it went into operation, searching several "nearby" galaxies for signals.

Still, no sign of another civilization was heard. But only a few radio frequencies were examined, and the antenna was aimed at only a few stars, so success really wasn't expected. In fact, all the searches made so far have looked at no more than 1/100,000,000,000,000,000th of

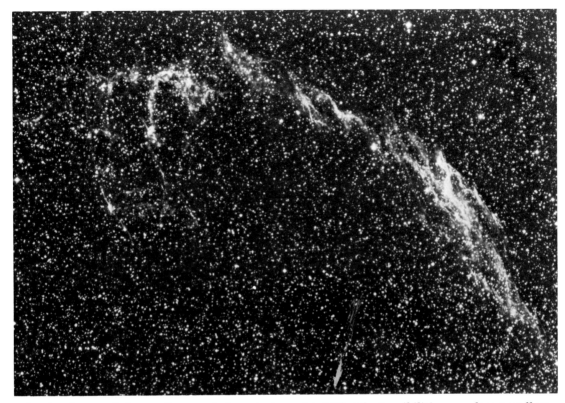

Scientists have only begun the search for extraterrestrial life. The possibilities are almost endless.

the possibilities—one out of one hundred thousand trillion!

At that rate, the search would take thousands of years. Obviously, a better way of doing it was badly needed. Using an electronic device known as a spectrum analyzer, astronomers can now search 10 million channels at a time. This device was used in both of NASA's major projects—Sky Survey and Target Search—in the early 1990s.

Sky Survey used a 112-foot (34-meter) dish to search the entire sky on frequencies from 1,000 MHz (megahertz, which is a unit of frequency) to 10,000 MHz. Target Search looked at 1,000 selected stars by using three radio telescopes ranging in size from 210 feet (64 meters) to 1,000 feet (305 meters). Although these were big projects, they covered only a tiny fraction of the possibilities, since most frequencies weren't checked. The two NASA projects were to continue for ten years but lost government funds and had to be put on hold. In the meantime, privately funded projects are expected to continue the search.

There is no way to estimate the chances of success for such projects since we don't have any idea where the signals are, or even if there are any signals to hear. After many years, we may have heard nothing at all, or we may get intelligible signals in the first five minutes of a new search. We may get signals that we don't realize are intelligent, since we don't know anything about extraterrestrial civilizations. In fact, this may have already happened.

Recognizing Alien Signals

One thing is certain: Signals from intelligent beings beyond the Earth won't be in English or French or any other familiar language. They may not even be in a language at all. Extraterrestrials may communicate with each other without words or symbols. If that is the case, understanding their signals will be extremely difficult, if not impossible.

Because language is how we communicate, that is what we will look for in the mass of static that comes from all parts of space. Meaningful information may be buried somewhere in the static. First we have to find it and recognize it for what it is. Then, we have to figure out what it means. That could take many years, unless the aliens have sent signals meant to be understood by humans.

If we are very lucky, they will have had previous experience with beings like us. They may be smart enough and experienced enough to know exactly how to get through to us. But if we pick up signals they send each other within their society (such as their equivalent of telephone conversations), understanding them will take the best minds we have. But it will be one of the most wonderful challenges that brilliant people have ever faced.

Could UFOs Be from Another Civilization?

It is just possible that while we have been looking and listening for any little sign of life outside the Earth, visible signs of life have been flying through our skies.

For more than 40 years, people all over the world have been reporting sightings of high-performance craft we call Unidentified Flying Objects, or UFOs. Many of these reports have come from airline and military pilots and others who aren't easily fooled by unusual sights in the sky. Because UFOs are said to look and to fly so differently from anything built on Earth, a lot of people think they may be spacecraft from some other planet.

While most UFO sightings turn out to be honest mistakes by people who really saw meteors, unusual airplanes, or very bright stars, many others are hard to explain. Also, there are more and more reports of strange craft seen on the ground, and of others that have crashed and been carted off by the government. None of these stories has been proven to be absolutely true, so the mystery remains. Groups of scientific-minded people continue to study UFO reports to find out what they mean. Perhaps some of them are alien spacecraft, or perhaps all of them are just mistakes.

Summary

All the major industrial countries on Earth have their eyes on space. They want to look and listen, probe and visit. There is something within the human spirit that will never rest while big, exciting questions remain to be answered. Just when humans will live on the Moon and walk on Mars has yet to be determined. But astronauts and cosmonauts will push back the boundaries of knowledge, bringing benefits still unknown to people on Earth.

For Further Reading

Asimov, Isaac. *How Did We Find Out about Outer Space?* New York: Walker and Co., 1977.

Bendick, Jeanne. *Space Travel.* New York: Franklin Watts, 1982.

Bergaust, Erik. *Colonizing Space.* New York: G.P. Putnam's Sons, 1978.

Berger, Melvin. *UFOs, ETs & Visitors from Space.* New York: G.P. Putnam's Sons, 1988.

Briggs, Carole S. *Women in Space: Reaching the Last Frontier.* Minneapolis: Lerner Publications Co., 1988.

Coombs, Charles. *Passage to Space: The Shuttle Transportation System.* New York: William Morrow and Co., 1979.

Deutsch, Keith. *Space Travel in Fact & Fiction.* New York: Franklin Watts, 1980.

Knight, David C. *Colonies in Orbit: The Coming Age of Human Settlements in Space.* New York: William Morrow and Co., 1977.

Mason, John. *Spacecraft Technology.* New York: The Bookwright Press, 1990.

Rickard, Graham. *Homes in Space.* Minneapolis: Lerner Publications Co., 1989.

Schulke, Flip and Debra, and Penelopee and Raymond McPhee. *Your Future in Space: The U.S. Space Camp Training Program.* New York: Crown Publishers, Inc., 1986.

Unstead, R.J. *A Space Station.* New York: Warwick Press, 1978.

White, Jack R. *Satellites of Today & Tomorrow.* New York: Dodd, Mead & Co., 1985.

Resources to Contact

American Astronautical Society, Inc.
6352 Rolling Mill Place
Suite 102
Springfield, VA 22152

American Institute of Aeronautics and
   Astronautics
370 L'Enfant Promenade, S.W.
Washington, DC 20024

Center for Aerospace Sciences
University of North Dakota
P.O. Box 8216, University Station
Grand Forks, ND 58202

The Commission on Science and Technology
U.S. House of Representatives
2321 Rayburn House Office Building
Washington, DC 20515

Eyes on Earth
146 Entrada Drive
Santa Monica, CA 90402

International Astronautical Federation
3-5 rue Mario Nikis
Paris Cedex 15 F-75738, France

John Fitzgerald Kennedy Space Center and
   Library
SAN 302-9905
Kennedy Space Center, FL 32899

Lunar and Planetary Institute
3303 NASA Road 1
Houston, TX 77058

The Senate Subcommittee on Science,
   Technology, and Space
Hart Senate Office Building Room 427
Washington, DC 20510

The Space Center
P.O. Box 533
Alamogordo, NM 88311-0533

Space Science Center
University of New Hampshire
DeMerritt Hall
Durham, NH 03824

U.S. Space Education Association
P.O. Box 1032
Weyburn, SK, Canada S4H 2L3

National Aeronautics and Space
   Administration
400 Maryland Avenue, S.W.
Washington, DC 20546

National Air and Space Museum
Smithsonian Institution
7th Street and Independence Avenue, S.W.
Washington, DC 20560

National Space Club
655 15th Street, N.W., No. 300
Washington, DC 20005

National Space Society
West Wing, Suite 203
600 Maryland Avenue, S.W.
Washington, DC 20024

The Planetary Society
65 North Catalina Avenue
Pasadena, CA 91106

Index

Acknowledgments

The photographs and illustrations in this book are reproduced through the courtesy of: Beong Huntsville, p. 27; Don Berliner, p. 12 (top); Boeing Aerospace, p. 23; British Aerospace Corporation, p. 14 (top); European Space Agency, p. 9; Martin Marietta, p. 6; MBB, p. 14 (bottom); McDonnell Douglas Space Systems Co., pp. 10, 17, 21, 57; National Aeronautics and Space Administration (NASA), pp. 8, 12 (bottom), 13, 15, 16, 20, 22 (both), 25 (both), 26, 30, 34 (both), 36, 38, 39, 41, 42, 43, 44, 46, 49 (both), 50, 52, 53, 55 (both), 56, 58, 61, 64, 65, 72, front and back covers; Rockwell International, p. 35.